GEOCENTRIC AND LOGIC

HUMAN LAWS ARE NEVER FOREVER
THINK ABOUT THEORY IS NOT A WEAKNESS

Have you ever tried thinking outside or defying a rule of law and trying to find something that you felt was better than the norm and what everyone recognized, and they all admitted that it was the truth? And the truth of what is the result of your thinking that you convey to others and it can be deciphered even if only in the form of theory which is based only on the visible eye and by observing the activities and authenticity that you see everyday. That is, I invite you to think even though it is seen or considered by many people as a folly and can make you called back to the primitive and hollow period of science.

That is why you will be seen as disrupting progress and underestimating the results of all kinds of research and all kinds of discoveries in various branches of science, which can now be utilized and have become the necessities of our daily lives. But do not you misinterpret, because I do not invite you to be a more stupid person or make you a person who is marginalized in the association. What I mean by dissent from the norm here is myself, so you don't have to feel like I have put myself in that position. So, the effects and consequences of my placement are certain, and only I myself will bear the consequences of what I do.

If there are people who ask what I mean, I clearly convey here that the law and custom - which in this case is man-made - is not something concrete and perfect. It means, the quality can still be improved and it might also be able to be removed and replaced with a new one which, if it is able to convince many people and be tested and accountable, will lead to our confidence in the new one. So I put forward the principle that I will describe as far as my knowledge of the matters that will be discussed. Because in essence someone who loves science will try to keep searching for the truth of something. Although maybe later what he conveys will be a subject of mockery and invite laughter and cynical views. But in reality, no matter how bad a person is, he must be looking for a way or something of truth in his life.

I honestly state here that I am nobody. This means I am not someone who has become a role model or a place to ask lots of people. Moreover, having the ability or sufficient knowledge about everything. But departing from the statement that something made by humans is not accurate and can not always be trusted through any proof, making me try to think and convey the results of that thought to others. Maybe through this writing what I thought would become something new and will become known to many people, which in the end caused those people to be disappointed with what they read about my discussion.

Maybe because I often read and also see natural phenomena, I was moved again to bring up the geocentric theory, which was followed by an oddity that is not something that is normal or easily accepted by most people today. But, I feel I have the freedom to think and think about it. So most of it and although not everything written in this book is directed or discussed about it. Which if reflecting on the statement earlier, then the contents of this book are more directed to the explanation and explain about the so-called possible contradictions rather than the laws and many theories that have been recognized so far and have become knowledge that is almost obligatory to be taught to students and students at high school and college.

In trying to describe what is my opinion, the most can or can only be discussed by someone about this knowledge without having adequate equipment and facilities such as me are only objects or objects that can be seen in plain sight. That is, the earth and other celestial bodies that are closest to us and can be witnessed with the naked eye are the sun, moon, and stars. With the naked eye or the naked eye we can see and assume or argue about what happened to the objects mentioned earlier and how the relationship between them. So when they are interconnected with each other we must discuss other objects when explaining one of the objects or phenomena that occur on that object. Obviously, if we discuss the earth, maybe we will also discuss the sun, moon, or stars in the discussion. What ultimately makes it happen to be a repetition of discussion with each other on each page of the title and specialization of the subject.

Humans are indeed given heart, reason and knowledge for their survival, and what has been achieved by humans today may be sufficient in the context of meeting the needs of life. But that, with the fact that happens that the more we know will make us more curious about everything. In essence, the higher a person's knowledge will make him more aware that more and more he does not or does not know. This is what makes it possible for everyone to think and learn and conduct research and trials to fulfill their thirst for science. But as humans, we also cannot ignore our limitations in terms of time lags.

So, what I write is a result or may be a benefit of my curiosity and the delivery of new ideas on a branch of science. That is, can be useful or cause problems and anger. And I also apologize for the limited ability to compose words, so that makes this book very thin because I can only discuss directly to the core problem. Hopefully the reader can understand.

Finally, gratitude to the Almighty is also the goal and starts what I do.

Sarik Taba, 29 Januari 2020

Author.

GEOCENTRIC AND HELIOSENTRIC

Since ancient times humans have tried to interpret what is happening in the environment and circumstances around it. With the advantages given sense and taste, humans have a desire and a sense of curiosity. Starting from that curiosity, people continue to try and try to find an end that means a meaning and a truth from that curiosity. Which although inserted by a little sense of pride or arrogance, but most people managed to find what he was looking for. However, even if the discovery raises a sense of pride or pride, it is clear that the results of the discovery in general are beneficial for human survival. And this proves that what is thought and produced by the mind is not something in vain. Truth is interpreted differently by each self, which in essence they all look for an essential and trusted truth. And that what is said is true is something that is recognized by many people.

Humans may be proud to have found something useful for ease in life, such as wheels, paper, machines, vehicles, medicines and so forth. But in the human body there is also a feeling called lust, which means desire. Yes, all wishes. How to make life easier, everything goes according to the will, and the desire to master. Although it is not easy to master the passions, some people get what they want to give themselves more value. However, a truth is sometimes not acceptable to others even though we convey. This is where a conflict arises, at least about knowledge about something that might only cause debate and not cause physical contact or armed warfare.

And also since ancient times, that man has raised his head to heaven. Not fast or too satisfied with what he knows on earth. But we also want to know about what's happening in all of nature. And the problem that we are discussing is about the debate about the center of nature, which is whether the sun or the earth. And this debate has also been going on for a long time and caused divisions even though it did not end in war. In this case it is about whether the center of the universe is the sun or the earth. Because by having each mind, humans can of course argue according to the range of their minds and how they judge things.

Geocentric, is one of the theories put forward and

states that all celestial bodies circulate or revolve around the earth on a line each of which continues to run regularly. With the earth stationary in place, it will make changes or differences in the appearance of the objects or objects that surround it. It is not a mistake for humans to think that way, because it is true that by witnessing what happens day after day. We watch the sun, moon and stars move regularly around us. This is called geocentricity, that the earth is the center, and other heavenly bodies revolve around the earth. Is this statement wrong? No, because at that time humans as a whole accepted and acknowledged that opinion.

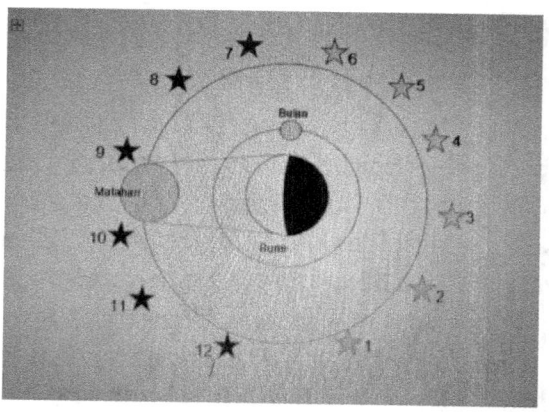

Image of Geocentric Theory.

But over time, and the world gives birth to new people who think differently who according to themselves is an intelligence. The notion of the earth as the center began to be refuted, and they put forward a new theory which was the opposite of the first theory. By stating that the earth and all heavenly bodies revolve around the sun, which is called heliocentricity. And this statement was raised and stated by those people as a truth. And amazingly, even though they did not actually see the evidence presented by the new inventors, they were immediately accepted by the public, which naturally changed the assumption so far that the earth was the center of everything.

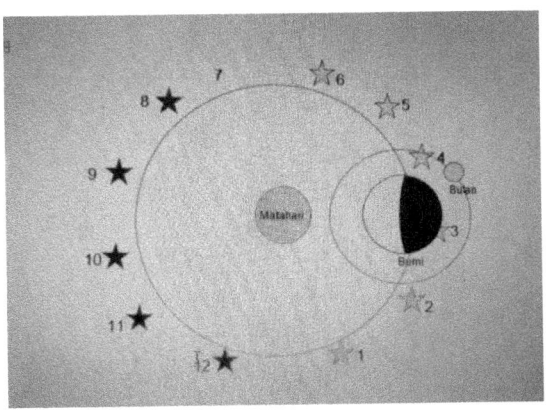

Image of Heliocentric Theory

At the beginning of this theory put forward, indeed led to an extraordinary debate. Especially with the religious leaders who made the new Tori adherents called heretics and forced to retract their words. Some even hanged or were exiled far away.

But by not extending the discussion, this book will also try to describe and convey what is in the head of the author of this booklet, which according to him is also a truth. The same as the descriptions at the beginning, it might just be a debate or become something that can not be accepted easily.

THE SHAPE OF THE EARTH

How or how much the shape of the earth we inhabit is a question and has become a subject of quite interesting and very popular debate lately. What is discussed or disputed is the shape of the earth as a whole. If the surface of the earth is clear we know that it consists of land and sea. Mountains, hills, mountains, oceans, sea trenches, straits and so on. This overall shape means what the shape of the earth is when viewed from a location that allows you to see the earth as a whole. And we know that one group thinks that the earth is round, and the other group states that the earth is flat or disk-like. They contradict each other and often argue with each other by providing evidence that is considered to be able to convince everyone of his opinion. No, it is not surprising that this matter is often debated, even as a matter of contention, of course the aim is to defeat, because it is related to several other matters which are also relevant and mutually reinforcing, including the comparison between the greatness of science and the science of religion and belief. Even this has become propaganda material to prove that a statement in the scriptures of a religion is more reliable than the scriptures of other religions, which of course directly makes the truth of their opinions as proof that their religion and beliefs are true.

But instead of diminishing the meaning of religion in human life, we will not discuss matters concerning religious matters in depth. Because this book is not something to be used as propaganda material and not for noise makers. What we discuss is what we see and we witness and, the writer himself certainly tries to make a logical statement and does not necessarily reinforce one of the opinions of the two groups. We will try to discuss how tested an opinion is and how opinion makers or a theory can maintain its hold. Even directly or indirectly this book may be a collection of theories or views of the author, although it may also be influenced by what is seen and read.

The round earth has indeed been a belief and hold for decades, or perhaps it has been centuries since the theory was first put forward. And indeed science and what is taught is indeed directing someone there. And not just talking, the adherents of the theory of a round earth like a ball or a marble are able to present concrete and credible evidence even though there are still many or some who still oppose it. Opponents, of course, adherents of the theory of flat earth.

The weakness of the round earth theory if it is studied and explored in science is not just one, many arguments and rebuttal can be submitted to oppose the statement about the roundness of this earth. And one of the most horrendous though maybe without reason and maybe also an option is a distrust of the study or investigation of science. Some people claim that all the evidence offered is not a pure, original or trusted thing. More briefly referred to as an engineering. But by not extending the subject, we will try to see. Yes, see. Because the only handle of the author of this book is to see. In a sense not assume according to what was heard and what was read.

The theory that states that the earth is round is more easily digested with reason. Because we can see the occurrence of regular day and night changes, sunrise, moon changes and so on. The three celestial bodies closest to us are the earth itself, the moon and the sun don't feel very different. We certainly can see that the sun is round and the moon is round. But when it comes to the issue of non-spilled sea water, river water flow, the matter of direction and flight time for airplanes for example, indeed the round earth theory is questionable. Regarding sea water that is not spilled for example, when compared to the parable of pouring water into a ball, the water will directly go down until the last drop falls under the ball and there will be nothing left. Could the earth if it was round in shape could hold so water, especially sea water, did not go anywhere? It is not wrong if someone asks about it. Then the river flow, if the earth is round why rivers that flow water that is looking for a lower place can climb a circle, that is also not a stupid question. And about airplane flights, why do two planes in different directions one go east and one go west with the same speed need the same time to travel the same distance. Supposedly if the earth were indeed round and spinning for example to the east, certainly one of the two airplanes would require a shorter time than the other planes. Is this a stupid question or rebuttal? Of course not, because the logic must be like that, because if the earth is round and spinning then an aircraft in the direction of the Earth's rotation will be slower to its destination or takes twice as long or more than the opposite of the Earth's rotation. That is, we can not arbitrarily draw conclusions and say groups or people who refute the theory of spherical and rotating earth theories are fools.

So, what we are discussing is why the earth is in the shape of a spinning circle and the elements that are attached to or connected to it can be what they are now, which means that they don't go anywhere. By seeing that every object that is tossed up or dropped it will point in the direction we call down. This proves that the earth has a power or force to attract something. That the earth will attract objects that are above its surface whose mass or weight can be maintained by the attraction of the earth itself. This means that if an object moves away from the earth, the earth will pull the object back to its original place. Thus, an object or object that wants to leave the earth must have a certain strength and weight in order to be able to resist the attraction. With the right weight, the right mass and energy, an object can move or leave the earth and escape even though it still cannot escape from the influence of the tensile data. Clearly, the earth is something that is already made and masters what is above it. So it is not the earth that adapts itself to that object, but if an object wants to move or move away, it must adjust or try to counteract the nature of the earth so that it can move or move at will. Which in certain objects must require human assistance in order to move or move.

Maybe in theory, the attraction of the earth is perpendicular from the earth's core to the objects that are around it. In a sense, if the earth is round then it will pull the objects above it perpendicularly, above in the sense of well below, beside, above or on its slope if it is likened to a ball. So the attraction will be the same for all objects that are around the skin without exception, according to the mass and specific gravity of the object. By considering the nature of an object is still, so if you want to move he must have an energy or energy to move. Perhaps with this theory humans are able to create motorized vehicles, airplanes and other artificial objects such as artificial satellites for example. In motor vehicles both two wheels, four and so on, the mass and weight of the vehicle must be adjusted to the power it has. With benchmarks that can be directly proportional, how much energy or energy is needed to be able to make a motorized vehicle move or walk, or how much weight and mass or weight of a vehicle to be able to move in accordance with the wishes compared to the energy possessed. Thus, the necessity for this driving force is absolutely necessary for an object to be able to move from its place, meaning that it must be able to meet a certain amount of energy or energy needs in order to move, because it has to adjust to the finished goods ie the earth with its appeal. This means that if compared to MotoGP, Valentino Rossi is a finished product. So, it is not Valentino Rossi who has to adjust to the motorcycle mount, but in order to obtain maximum results or in accordance with what is desired then that motorbike must be set in such a way as to match and match the weight and style of racing Valentino Rossi. Likewise in airplanes, with a certain weight, mass and energy, the airplane can move over the air after it has been set up so that it can be controlled after adjusting to the earth's attraction.

The current level of advancement and sophistication of science is indeed quite extraordinary, and we can see and witness with clear evidence that something humans obtain from their knowledge. We can say one of them is the existence of artificial satellites. How satellites can stay in place - meaning that they are always perpendicular to the initial attraction of the earth where they were placed - without requiring any more energy after being launched. After being placed in the right location, the satellite will move regularly and automatically or it might be more correct to call it still in a place that is pulled by the attraction of the earth. By giving an example of the Palapa Satellite in Indonesia, he is always in his place and follows wherever Indonesia rotates, while there is no damage or disturbance. This proves that we can determine the mass and weight of something in order to adjust to the pull of the earth. So, do not be surprised why sea water, rivers and others can be like now if objects that are far from the earth still have to adjust to the earth. So in conclusion, the earth is finished goods and other objects must adapt or must follow the wishes of the earth, or humans must do something in order to be able to do something too in order to adjust or escape from the influence of the earth.

About the theory of flat earth is not impossible if we put forward such assumptions, water, sunrise and sunset, and ice freezing on earth. By proposing a theory of how the sun rises and sets and so on. We can hear or have heard that the earth is flat and the sun circulates periodically and regularly by going up on the east side and down to the west, then sinking which means passing under the earth's disk to come out or rise again the next day on the east side. Or there are other opinions of the sun such as a flashlight and moving regularly also right in the middle, meaning that it runs right between the midpoint of the earth and the outer lines of the earth. With movements like this will result in the middle of the earth and the outer side will get less sunlight. With the lack of light which means no heat will cause ice clots to form right in the middle of the disk of the earth and around the outer edge of the earth. Specifically this outermost part will cause someone who surrounds it if moving from one point will turn continuously in one direction, meaning we will circle the earth on the edge of the outermost line by taking a clockwise direction so we will surround the outer side of the earth with the direction always right to the wall ice clots that serve as a barrier are always on the same side.

The two theories about the flat earth above may be true, but the main weakness is to pay attention to the nature of the sun's light that spreads. The first theory is that the earth is flat and the sun rises in the east and then sinks in the west, as if this were impossible. Because if the earth is flat, and the sun rises in the east then the whole plain will experience the time of the sun's rise or rise at the same time. That is, the entire surface of the earth will experience lighting by sunlight at the same time. But the reality that happened and we witnessed was not like that. What we are witnessing is that the parts of the earth's surface have a continuous, gradual and regular lighting time to get sunlight. So that the greater the distance of a location with other locations, it will cause the greater time difference between the two locations that it will be lighting by the sun.

Indeed, this theory states that the sun has a path that encircles the earth with loops up and down making it possible to rise and set, and it may be that the intensity and level of illumination is different when viewed from the angle of its bright quality. That is, the east will be brighter than the west, which gradually dims due to the farther distance. But still, if the light is fired or directed at a flat plane then the entire surface of the plane will get the same light at once, and maybe the only cause of not being accompanied by light is simply because there is a surface that is not at the same height or protected by hills or mountains. Clearly, the base of the east and the west end will see the light at the same time or at the same time even though the quality of the information and the temperature are different due to differences in the distance from the light source and the heat source.

Another objection is to pay attention to the surface of the earth itself. Which consists of land and sea. If indeed the sunlight crosses right in the middle, let's say the latitude of 0 degrees, then it will be increasingly apparent that the theory is contrary to the nature of nature. As we have seen, the sun is almost all year around the latitude of 0 degrees. And we can prove that the entire surface of the earth that is around or along the equator is not found ice or even just snow for example. If the sun's circulation is like that, then it will cause water flow will only occur around or along the line, with the ice walls to the left and right whatever the height of the ice wall. And back to running to the nature of water, there will be no dividing wall blocking the water in the midline to form a large collection. The water will spill up to east or west, which will make us never know the name of the sea or ocean.

The second theory of flat earth is that the sun is like a flashlight illuminating the earth. With the line in the middle of the radius of the earth's circle. The sun goes along the line by firing light focused on the passage. With a fixed shift, then the parts will get rations alternately with a certain time as well. Which certainly translates to 12 hours for each region or area. And this could have happened had the sun also been in the shape of a luminous plate and its rays were focused on a field such as a laser beam or a flashlight, because if the light spread it would still be visible from any region throughout the plain. This theory is also reinforced by the statement that the center of the earth or the midpoint of the earth is the area we call now the north pole. With a clear explanation that the ice at the center of the radius of the earth is caused by lack of getting sunlight, while the outer part which is now called the south pole also freezes due to the same thing as what happened with the midpoint. So the so-called equator is the path or path of the sun in the middle of the radius of the earth, while the north pole is the center point, while the south pole is the area around the circle. An amazing theory.

But unfortunately, if the sun is in the form of a plate or a luminous disk that is in focus, there will certainly be a change in the shape or appearance of the disk of light. Or even the sun may be rounded with focused light, there will still be a change in the appearance of it. This will cause the sun to look thin at sunrise, then expand into a circle in the middle of the day. And of course it will return to its original form of thinning again until it sinks or disappears. While the reality that we see is that the sun always has the same shape from sunrise to sunset. And if translated further it will cause the sun to appear to move in a circle to the side and not in a straight line even though we are right in the middle of the radius or the so-called equator.

The biggest weakness of adherents of the theory of flat earth is the absence of pictures or photographs that can be carried as evidence of that statement, or maybe never conducted an investigation about it. So what is conveyed as if it was only fabricated without proof. But maybe because human nature that always wants to win makes adherents of the theory of flat earth always looking for ways to prove the truth of their opinions. At the same time we hope that hopefully in the future we can get concrete and concrete evidence that the earth is flat as they say.

EARTH ROTATION

There is no doubt that the rotation of the earth for spherical earth is a must and a thing that has become a role model for everyone, about how the earth rotates on its axis and what are the effects and effects of that rotation. The earth rotates on its axis, this axis is called stretching from north to south which results in a change or change in position between east and west. With the rotation of the earth will result in some real natural phenomena that can be witnessed by everyone everyday. Regular day shifts and day and night changes throughout the earth's surface. Assuming that the earth has been rotating on its axis for 24 hours in one day, what is evident in the area or region that is closest to or in the earth's equator is that it changes constantly, regularly and measured throughout the year.

At school, it is taught that the earth rotates on its axis with an oblique position of several degrees.

That is, the axis of the earth is not perpendicular between north and south, but rather leaning forward or sideways to a number of degrees. With also the statement that the sun is still in place makes the time (units of minutes or hours) can be divided into several parts. And what is used today is that the world is divided into 24 time zones by referring to 0 degree longitude set in the UK, so that the determination of the time in a place is adjusted to where a place or region is located or what is the degree of difference from the 0 degree longitude. And it is true that the division of time can be said to be accurate and sustainable. By dividing the longitude of a circle in general or used is equal to 360 degrees. That is, the earth's circumference between east and west starts from the point of 0 degrees to 360 degrees if divided 24 hours is 15 degrees each time zone. With the breakdown of one time the earth will cause the time difference between each region in different longitude. With current reference, West Indonesia Time (WIB) is different by about 105 degrees from 0 degrees longitude making a difference of 7 hours from the area along the 0 degree longitude. Because the earth's rotation is from west to east, making WIB 7 hours faster than the 0 degree longitude. Ie, if the WIB has shown at 07.00 in the morning then the area at 0 degrees longitude is still at 00.00 midnight.

The relationship of rotation of the earth may be related to how the formation of the earth, some theories that state how the formation of the earth is the Big Bang Theory and the formation of celestial bodies from fogs and nebulae, although there are many other theories, but the two theories until now have not been proven concrete. But because the subject of this book is the universe, these two theories are what I have mentioned to be discussed. Both are only limited to theory, which because there is no other theory makes them both so famous and most often discussed. But I say that we will not discuss the two theories. Both of them I mentioned as a theory that has not been proven let alone tested.

The different views on the relationship between heavenly bodies and the earth, make statements that are also different or contradictory. Because one declares the sun as the center, while I am on the other side which states the earth as the center. Regarding rotation, the earth is stated to rotate on its axis along 360 degrees in one day or 24 hours regularly throughout the day and year. But because the point of view is not the same, it makes me say that the reality is not like that.

Now you have to look up at the sky, do you see the change in the location of the sun, moon or stars? The average of the three celestial bodies is seen for 12 hours at a location on earth. The sun rises and sets, so does the moon and stars. Then what is the relationship between the earth and the sun? In fact, the initial yardstick for writing this book is how to prove that the earth remains in its place. And all heavenly bodies including the sun, moon and stars circulate around the earth. How is the explanation?

The sun is actually the same as the moon. The only difference is that the sun has its own light while the moon does not. The equation is that both evolved around the earth. Indeed there is such a thing as pseudo daily motion, indeed pseudo, but the understanding does not stop there. The daily apparent motion of the sun as if walking from east to west. Rising in the east and seen for 12 hours then sinking in the western horizon, after sinking came the night for 12 hours (at the equator) to return to wait for the sun to rise the next morning.

Now, if one year is 360 days and the sun's circle is 360 degrees, the earth actually rotates on its axis 360 degrees plus 1 degree, meaning 361 degrees every day for 24 hours. The relationship with the sun is that the sun is indeed like the moon evolving around the earth, but at a much slower speed. Both are moving from west to east at their respective speeds. If we assume that the circumference of a circle is 360 degrees, then the sun shifts by 1 degree per day to the east. This is certainly a question, if the moon looks at the place and time of rising always changes every night, why does the sun still rise at the same place or at the same time every day.

This is caused by - not yet calculating the tilt or shifting of the sunrise position from north to south or vice versa - the movement of the sun from west to east by an excess of degrees of rotation of the earth rotates for one day. With the rotation of the earth at 361 degrees, it means that the sun will shift eastward by the excess. If likened to 0 degrees longitude on the first day is at the starting point that is 0, it will move east by 1 degree. To be exact, the earth takes 24 hours to rotate 361 degrees, while the sun needs that much time only to shift by 1 degree. So, if we assume the starting point of the solar revolution is at 0 degrees longitude, then the sun will continue to follow the 0 degrees longitude so that it can rise in the same place the next day. That is, 24 hours to circle 361 degrees of rotation of the earth is equal to 1 degree shift in the location of the sun to the east. That is, an excess of 1 degree of rotation of the earth with a count of one year will put the point of 0 degrees back in its original position against other heavenly bodies. And in Indonesia, for example, with an appropriate ratio between the excess degree of the earth's rotation and the speed of the solar revolution, it will make if the sun rises at 6:00 a.m. it will remain constant throughout the year due to the accuracy of the excess rotation with the speed or distance of the sun.

As a result of the rotation of the earth maybe everyone already knows, but what causes the rotation of the earth has not been much discussed, even unheard of at all. Here I will try to explain that. We can only see that the movements of the sun, moon and stars do support this. This means that there must be something that causes it, or what makes it that way. And here what is being challenged is your belief in what governs the entire contents of nature. Almighty. However, in the context of science that promotes freedom of thought and logic, maybe not everyone believes that everything that happens in nature is solely with the words of the Creator with Kun Fayakun's words, Be Happen, Then Happen. As religious people we certainly have to believe in the power of the Creator, but the process of creation we do not know whether at once or through certain stages that require a certain amount of time.

As mentioned above, the origin of the universe is stated by two well-known theories, namely the formation of the universe from the big bang or the Big Bang, and the formation of objects in the universe from dust or nebula rotating in a nucleus. In the Big Bang theory, a large explosion makes an extraordinarily large celestial body explode, resulting in an infinite number of fragments. But, if the explosion is true, why should there be a core or center of the explosion because all celestial bodies move in a circle. If it is true that the explosion should move all celestial bodies away from the center of the explosion, each of which takes care of himself without regard to the other because there is no frictional force or brake alias from the movement of the object. That is, there is the title universe expands. But as we know the reality or theory that there is a celestial body rotating around a center or parent or queen bee with a fixed speed and direction.

The second theory may make more sense, namely the formation of an outer space object from dust or nebulae that move around a core or center. The longer the dust becomes more and more solid and more dense until a celestial body rotates. But the weakness is that we just model our earth, still spinning but not enlarging, even the atmosphere or ozone layer is said to be getting thinner, rather than getting thicker.

I mentioned those two theories only for how to describe the rotation or rotation of the earth. As we know, the earth has a magnetic field. If we have been mentioning the magnetic north pole and the south pole, in fact it is wrong. Because the direction of magnetic attraction is east and west. With opposite poles, east and west, it is possible for magnets to push each other together and together try to attract objects in the form of magnets or other magnetic metals that are above them. That is, if a magnet or a metal that is near the rotating crust will affect the magnetic core. But the problem logically is who first moved it.

Another possibility is if you think everything is from zero and back to zero. So it can be said that the earth before the Prophet Adam was sent to earth, the earth has been perfect in its condition or the process has ended. That is, the earth that did not exist before existed whether you held to the Big Bang or the Nebula theory. The process of the occurrence or creation of the earth is certainly gradual, so that it becomes what it is now. Starting from the formation of the earth's core, the creation of layers and surface of the earth, to the growth of plants and the formation of the atmosphere. In the process of creating it, the earth always experiences an increase and change in itself. The earth that we encounter now is after going through a process of creation or creation. The constituent components gradually form and become the composition of an object called the earth. All components, both magnets, rocks, iron or soil and plants, mountains, fire and water. After all the components are complete and neatly arranged in such a way as to make an object that stands alone. In the process the earth that did not exist came into being then gradually increased to become a complete object. Like making a machine or a car for example, when all components are complete and the engine or car can operate, then the earth becomes neatly arranged to be able to operate or rotate.

So with that assumption or theory, as we demonstrated in the car or engine, when the car has finished being made, and is ready to run which is supported by the right components. However, after being used, then the function of each component will gradually change or more accurately be said to be damaged. Shiny skin or surface will fade from the

GEOCENTRIC AND LOGIC

sun and rain or scratch, worn tires, or the engine system will no longer be perfect. Which eventually will make the car occasionally have to enter the garage or service. But however we take care and fix it, the car continues to age and eventually becomes junkyard and rust.

That way, when we return to discussing the earth's perfect composition and supporting material, if disturbed or damaged, even though we call progress for human civilization. It turned out that the discovery and most of the mining of the bowels of the earth, and also the disruption that occurred in the earth's atmosphere that might thin out due to our own actions will make the earth damaged. And as we know an item that is damaged will be reduced in function and no longer stable as when he was first there. What I am sure of on earth is that the existence of a large magnetic field function will no longer be balanced as we continue to explore and retrieve the supporting elements of the rotation. Which is because he is like a body, if one element and function of the element is reduced it will make it weakened or damaged or sick.

With this view, it is not impossible that at some time the earth will gradually experience damage or will be reduced or damaged one of its components or elements, in terms of the elements that make up the earth's body that can be called a counterweight to that rotation. With the reduction or destruction of one or many of these elements or components will result in instability or disappearance of the counterweight from the rotation itself will make the earth's rotation slow down or even stop altogether or even reverse direction.

SUN REVOLUTION

Have you ever thought that the sun is similar to the moon, in this case its movements? As mentioned at the outset, this book is trying to find contradictions in science, especially in astronomy. Of course the above title looks ridiculous or too extraordinary. But there is a sentence that states "an extraordinary claim requires extraordinary evidence". Departing from that sentence makes me believe that if we make a statement that is out of the ordinary, may be delivered or announced or claim at the same time offering evidence from that statement. As an expression that a theory or a law is not concrete and perfect without proof. And indeed, thinking clearly in tranquility will make us feel comfortable and find an inner calling about whether or not something is necessary. Thinking within reasonable limits is normal for us humans, especially about new things that we feel are better will make us more motivated and challenged to do something for the sake of truth.

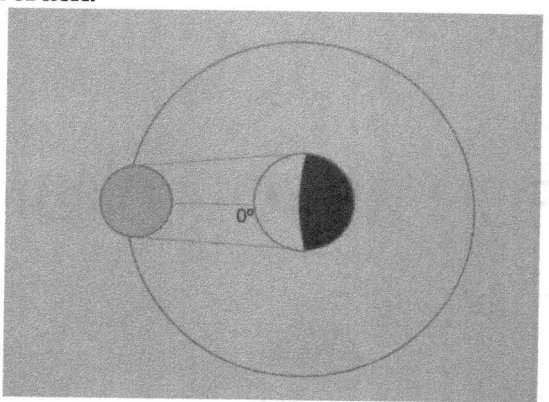

Drawing at a glance sketch of the solar revolution

Returning to the past is not a mistake if it turns out that the past is better and more useful for us. Technology that is now considered extraordinary does not make all our knowledge about something has been perfect. Everything returns to the initial belief, if we think that this or that is true then we will step up and hold on to the core and core of our thoughts. The goal, of course, is still in the context of looking for more evidence and arguments that can be put forward that what we make the basis for thinking is true. But, instead I underestimate and assume that all knowledge of the solar system and galaxies that are considered to have been very advanced, understand that as I say that we walk in the direction in accordance with the basis we think. To find proof. Because today everyone thinks and is based on the belief that the sun is the center of the solar system, so everyone will think about how to find more advanced evidence about it. But because I think with the opposite basis, then I also try to find proof of what is the basis for that thought.

The sun around the earth does not seem too strange a statement. Because in the past, humans have thought that way and become a belief for hundreds of years. However, because there are other statements that are considered to be more understandable and trustworthy, the assumption is abandoned. Even though it may have been hundreds of years that the sun is the center of the solar system or maybe the center of our galaxy, making statements that contradict it look as strange and disturbing as it used to be that the earth was not the center of everything in the universe. The sun surrounds the earth. That is the basis for my thinking. And will try to explain as much knowledge as I have about it.

As mentioned earlier, because four main objects: earth, sun, moon, and star, are interconnected with each other if you may be considered in living their lives. And as stated that if we discuss one object it may also have to discuss other objects in the specialization of the discussion. Which might make it discussed repeatedly. So in the discussion of the sun around the earth, of course we will discuss what to do with the earth itself.

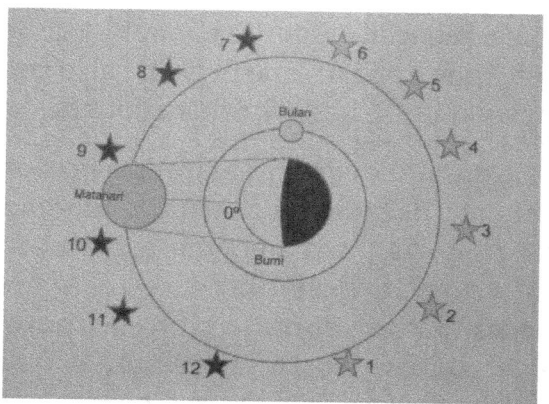

Graphical arrangement of the Sun, Earth, Moon and Stars.

As our knowledge now that the earth around the sun for 365.25 days or 1 year. Finally, we will continue to try to find evidence about it. However, if we reverse the fact that it is not the earth around the sun but the sun that evolves around the earth there won't be too many striking differences. Including its influence on seasonal changes, the change of years, and the division of regions or time zones for example. If we can accept the statement that the moon orbits the earth why don't we try to think and accept the fact that the sun is the same. Because as we see everyday, the reality of the movement of the two objects is the same. Rising in the east, sinking in the west. And in theory a rotating spherical earth feels it is not impossible that both have the same nature or movement towards the earth.

Our belief so far is that the earth does not stand perpendicular to its two poles. But rather tilted at 23.5 degrees forward or sideways. This book will also discuss the slope and its impact on life on earth. And I clearly state that on the basis of thinking that it is not the earth around the sun but on the contrary the sun rotates around the earth, making me think that it is not the earth that is slanted, but the sun's orbit which has a slope. The effect or effect of the slope of the sun's path is the same as the effect caused by the earth's theory around the sun in an oblique position. And about this we will also discuss at length in a specific title.

The earth rotates on its axis for 24 hours, with a circumference of 360 degrees. And the general sunrise at the equator is seen for 12 hours every day. This happens because the sun only moves 1 of its 360 degrees in 24 hours. Indeed we see the sun - at the equator - always being published and published at the same time every day and throughout the year. Then why can the sun come out when the moon around the earth is always changing every day?

This is because the earth rotates around its axis not 360 degrees in one day or 24 hours, but the earth rotates or rotates on its axis more 0.9856 degrees every day, which is to be 360.9856 degrees. The relationship with the sun is so that the sun still rises in the same place (in this case longitude at the equator) then the sun must move by 0.9856 degrees in 24 hours. This 0.9856 degree figure is obtained from, if one circle around the circle is 360 degrees while the time needed by the sun to evolve is 365.25 days then that number will be obtained. With details 360 / 365.25 = 0.9856. This if we take only up to tenths of a thousand numbers. If we add or increase it to a million or tenths of million, the number will be more accurate. But we will only use up to four decimal places or tenths of thousands to avoid taking up too much space.

So when we take the example of 0 degrees longitude moving after one day at 0.9856 degrees to the east, then the sun is already aligned with the 0 degree line the next day. This means that the sun moves or runs in a circular position as much as the excess degree of rotation of the earth, whose movements are certainly very slow because it is only 1 / 365.25 multiplied by 360 degrees every day. We will not play around with numbers to determine how much the distance of the sun in a day is in units of distance or kilometers, because that requires accuracy of how much the distance of the sun from the earth. The-sorry-is still fabricated.

Now suppose 1 year is 360 days, then the sun needs 360 days to complete 1 revolution. So when the solar revolution is 360 days, without causing disruption to the rotation of the earth. If we only measure the rotation of the earth and the sun then this discussion might have been quite understandable. In the sense that the sun will always be exactly at the longitude where it originally evolved because it always follows the amount of longitude shifting where it originally evolved. This is what causes the sun to always be seen rising and setting at a fixed time or time in each area of time or longitude, especially those that are right on the earth's equator.

But as stated that every discussion of an object will be related to other objects, then the solar revolution and the rotation of the earth as such will also determine the changes in the appearance of the moon, changes in constellations, changes in seasons, and possibly also eclipses. However, with the specialization of the discussion, the effects or consequences of the relationship between the sun and the earth will be described individually in a special discussion.

Besides having a speed in evolving, the sun also has a slope towards the axis of the earth or the equator. All this time we have argued that the earth is tilted but in reality the sun has a slanted path to the earth. The tilt of the earth that people believe is now at 23.5 degrees. But if measured by taking a reference point will try to test how accurate the estimate or the results of the study. By taking a reference point or a location on earth we will find out what the actual degree of slope is.

In circulating around the earth, the sun is not always parallel to the equator, although the numbers

I offer are not accurate but with full consideration and calculation I say that the slope exceeds 28 degrees. No need to be too complicated in formulating about it as well as proof.

As we see now, the sun is right at the equator twice a year. Namely on March 21 and September 23. To prove the magnitude of the slope, we must take a reference location. And the general public basically knows that the location that I made a reference to is not a foreign location for everyone, the location of that reference is the City of Mecca or rather the Kaaba. The reason for making Mecca or Kaaba as a reference point is because Mecca is at a latitude of 21 degrees, and also as we both know that Mecca is right under the sun or the Kaaba has no shadow occurs 2 times in one year, to be exact on May 28 and July 15. Means that Mecca is traversed by the sun's path exactly on these dates can be calculated:
March 21 = 0 degrees
May 28 = 21 degrees
So the distance is for 67 days. So the calculations are:
21 degrees divided 67 days = 0.3134 degrees.
Can this calculation not be used as a reference to calculate the magnitude of the slope change?

By calculating or with the determination of the solstice of the sun on June 22, it takes 93 days if calculated from the latitude of 0 degrees (equator) on March 21. So with a shift or change in the location of the slope, which averages 0.3134 / day, the northern

turning point of the sun is:
0.3134 degrees x 93 days = 29.1462 degrees.

However, with the possibility or certainty that the peak of a circle if surrounded within a certain period is half of the peak, the count or the peak or solstice is: 29.1462 degrees minus 1/2 of 0.3134 (the result is 0, 1567) is 29.1462 minus 0.1567 = 28.9895 degrees.

With further proof, Mecca is right under the sun on July 17th. If again calculated from May 28 to June 22 is 26 days, then by counting June 22 to July 17 is the same, which is a number of 26 days. Next, please count July 17 to September 23 where the sun has returned to the latitude of 0 degrees or the equator. Then the amount obtained is the same as the time required from March 21 to May 28, which is 67 days. This is not a coincidence.

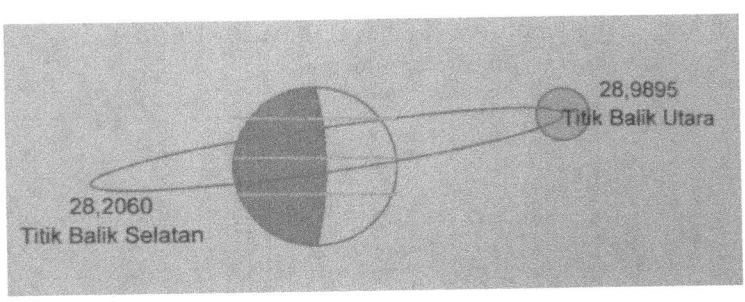

Image of the Slope of the Solar Revolution against the Equator

By looking at the figures above, we might wonder why the sun is 186 days in the northern hemisphere.

With one year 365.25 days or 366 days there will only be 179.25 or 180 days to the south of the equator. We should not be surprised, with the speed of the sun that is constant or the same every day, this proves that the solar path is not always the same distance to the earth, or the earth is not right in the middle of the circle. So, the earth is located slightly south of the center of the circle it should be.

The speed or distance of the solar revolution will certainly greatly affect other things, such as the change of seasons, the appearance of the moon, or changes in the appearance of constellations and eclipses. What is clear is that the above description is visible to prove that the sun's slope is not 23.5 degrees. Which although all this time is called the tilt of the Earth's axis. But this book is the opposite as well as the rebuttal of the theories that have been holding so far which hopefully does not make it something that will unsettle many people.

GEOCENTRIC AND LOGIC

Sketch Drawings Travel and Position of the Sun in evolving.

REVOLUTION OF THE MONTH

The moon evolves around the earth, this is nothing new. This has become everyone's guideline in its application in science. With a 29.5 day revolution, the moon regularly moves throughout the day and throughout the year. By experiencing changes in face or appearance of surface area that occurs every day. The moon dies, sickle, full moon, and dies again. Obviously, the moon's revolution against the earth is a fact that has become a law in astronomy.

If you read the paragraph above, it seems there is nothing else to discuss. Because everyone already knows. But there is a slight difference about our understanding of the moon so far that needs to be explained. This explanation is needed because in the previous discussion we have discussed how the solar revolution. Which in itself would confuse the current assumptions about the movement of the moon and its relationship in terms of position and appearance. This chaos is caused by changing focal points. As we have discussed about the factual distortion between the relationship of the earth and the sun, it will certainly also greatly influence the movement of the moon. Although it remains in the fact that it surrounds the earth, there has been a slight change in the rules that must be followed by the moon to keep it looking as usual or as we see it every day.

The moon and the sun evolved around the earth in the same direction. What distinguishes it is, the moon is closer than the sun to the earth. So they both form a circle in the form of a circle but the magnitude of the circle is not the same. Closer to earth, making the moon's orbit smaller than the solar orbit. Another difference is the speed of evolution, which if adjusted to the view that an object closer to the earth will be more affected by the movement and pull of the earth itself, which makes the moon that is closer become faster attracted or forced to circle the earth faster in accordance with Earth's rotation speed. With longer distances, the sun will move slower because its weight or mass is affected less than the effect of the Earth's rotation on the moon. Which causes the speed of the moon to evolve - in this case the number in degrees - than the sun. The difference in speed is what determines changes in the appearance of the moon or the processes that cause eclipses, both solar eclipses and lunar eclipses.

Under normal conditions, to complete one revolution against the earth, the moon when the revolution is 29.5 days should move at 12.2034 degrees per day (360 / 29.5 = 12.2034). With a number of degrees of 12,2034 this month will complete its revolution in 29.5 days. But with the discussion of how the sun's relationship with the earth, made a slight change in the revolution of the moon. As already explained that the earth rotates with an excess of 0.9856 degrees and the sun surrounds the earth at a speed whose magnitude is balanced with the excess rotation of the earth which is 0.9856 degrees per day. If we hold on to the current opinion that the change in the appearance of the moon is determined by its position on the silent sun, it will certainly result in a change in appearance that occurs on the moon will change or be chaotic. Because all this time we assume that the sun stays in its place. So with the handle that the sun evolved around the earth, something must also change in the principles and rules of the moon's revolution.

With the change in the location of the earth's longitude and also the location of the sun as described, resulting in the moon must adjust to it. So the moon, which should move only 12.2034 degrees per day, must increase its speed. This is required because for example we take longitude 0 degrees earth which is always parallel after one rotation with the sun, always shifting at 0.9856 degrees to the east. So the change in the location of the longitude of 0 degrees is at once the location of the sun perpendicular to it during the moon revolution of 29.5 days is: 29.5 x 0.9856 degrees = 29.0752 degrees. So for the moon to still look like it is now with regular changes in appearance and appearance, the moon must pursue the 0 degree longitude in its evolution. The point is that the moon that should have evolved by 360 degrees was forced to catch up to 29.0752 degrees. This makes the revolution of the moon amounted to 389.0752 degrees which was taken within 29.5 days. With the time of the revolution remained but with a different speed. With 389.0752 degrees per 29.5 days, the speed of the moon per day is 13.1890 degrees. This has to happen on the moon to look as we see it as usual.

Changes in the degree or position between the moon and the sun is what determines the appearance and appearance of the moon, the sun will illuminate the moon with a beam angle according to the magnitude of the difference. To prove this we need proof, which can be deciphered with numbers. Let's look at the following example:

Day 1
Sun   = 0.9856
Month   = 13.1890
Difference   = 12,2034

Day 2

The sun	= 1.9712
Month	= 26.3780
Difference	= 24,4068

7th day
The sun	= 6,8992
Month	= 92.3230
Difference	= 85.4238

Until here we can see that the difference from the movement of the sun with the movement of the moon is the same as the distance of the moon that should be if the sun is not moving. This is in accordance with the change in location which affects the angle of sun exposure to the moon visible from the earth. And from the example above we can also see how many degrees the height of the moon when it starts to appear or at the beginning of the visible month after the dead month which marks the arrival of a new moon in the calculation of the Islamic year. With a revolution of 29.5 days at the equator, the moon sometimes dies at 6:00 in the morning or at 18:00 in the afternoon. This means that if the moon dies at 06.00 am, the moon will rise and look as high as 6.1017 degrees in the western horizon at dusk at sunset or we call the crescent. Likewise, if the moon dies at 18.00 in the afternoon it will be seen as high as 12.2034 degrees in the western horizon the next day.

For example, the 7th day is to show the degree of the moon on that day. With a difference of 85.4238 degrees at the time of publication, as we know in the Hijri calendar called the rising moon, the moon changes at any time, meaning that if it rises at an altitude of 85.4238 degrees it will sink at 85.4238 degrees plus half the day trips of 6.1017 degrees will be 91.5255 degrees. In this condition or position the moon appears right above our heads if we are at the equator. With a difference in the degree of irradiation of 91.5255 degrees we will see half the surface of the moon from the earth. Like a plate halved. Because of the earth's rotation, the moon will rise at 18:00 in the afternoon and sink at 00.00 at midnight.

On the 15th day or night the moon is at an angle or height of 197.8350 degrees when it rises when viewed from the earth. While the sun is at an angle of 13.7984 degrees. There is a difference of 183.0510 when the moon rises. With the magnitude of this irradiation angle which means the location of the sun and moon facing each other right on the earth, the moon looks full as a circle. This is what is called the time of the night or full moon. The moon rises right at dusk with a round surface on the eastern horizon and is seen all night until sunset the next morning at 6:00 in the morning.

The lunar revolution as we witnessed is that at the beginning of the Islamic month the moon will look faster and the sunset will also be fast from day to day until the middle of the month. And will be published more slowly with sunset time which is also slower until the end of the month. We can prove the late sunset by still seeing the moon even though it appears to fade during the day when the sun is still shining.

Like the sun, the moon is not always right in line with the equator. The moon also has a slope towards the point or core of the earth. Perhaps scientists, through continuous and ongoing investigations that have been carried out for so long, have succeeded in finding out what the degree of slope of the moon's orbit is. The slope of this path can be proven by not always rising and setting the moon right at the equator, but shifting north or south. The speed of the sun and the moon in its evolution and its tilt on the earth is very influential on the occurrence of solar eclipses and lunar eclipses. Although the author has not received exact figures, it seems that the tilt angle of the moon is greater than the tilt of the sun. This conclusion is drawn from the non-always occurring lunar eclipse every full moon, and the solar eclipse does not always occur every month to die. With the reason that the magnitude of the moon's slope is smaller, the likelihood of a solar eclipse will be greater every dead month or new moon, because the speed of the moon in evolving is almost 12 times the movement of the sun. So that if the moon only dwells on a small slope with a very close distance to the earth and has a very large difference from the distance of the earth and the sun, we can be sure there will be a solar eclipse even though it is only a partial solar eclipse.

ASTROLOGICAL SIGN

Constellation, is a very popular term. Besides being useful as determining the direction and counting of months and years, constellations in certain countries or countries are also used as calculation material to predict someone's fate. Although it is not fully proven and only becomes a clue that is not in accordance with logic if it is connected with one's destiny and way of life. But that does not reduce the interest of many people making constellations a sign that was called fortune telling. The constellations that are usually referred to are 12 types with each name. According to their belief the zodiac is a material to guess what will happen today and will come, and also affect one's life path.

Constellations are formed from the arrangement of several stars that form normally a series which when connected with several lines will form a pattern. This pattern is named after the shape of the animal or creature or object that is almost like it. We certainly know the names of constellations such as Aries, Pisces, Big Dipper, Scorpion, and others. Giving this name as conveyed is adjusted to the shape or pattern that occurs from the arrangement of the stars. In the science of logic and free mind, belief in the science of the zodiac may be far removed. Despite the fact that the constellations are seen regularly and alternately periodically on the horizon of the night sky. And indeed, we will see changes in star formation that occur gradually each night and every month.

Changes in constellations as believed so far have been caused by changes in the position of the earth in its journey around the sun. This opinion is not unreasonable because it is easy to understand. We certainly will see objects or objects that are different if we rotate our view direction. The things we see in front, back, and side of us will not be the same when we stand in the wild. Like that, too, what happens to our views if we raise it to the sky. We will see different stars throughout the year because we walk in a circle or around and return to the starting point of the journey according to the theory of evolved earth.

Now if we show that there are 12 stars, 1, 2,3, 4 and so on. If we walk on the star in a circle then we will see star 1 at the beginning of the journey and continue to advance until seen until the 12th star after one rotation, and will return to see star 1 when returning to the starting point of our journey.

The sun is also called a star, even though its size is much bigger than other stars because of its proximity to us. Compared to distant stars of course the sun looks bigger. So it can be said that all the stars are behind the sun if we look at this from our earth. This description would have become everyone's daily food. Because since elementary school we have learned and taught about it. However, because this book is the truth of the knowledge taught in these schools, the description and explanation of changing positions and constellations will be different from what we are used to.

The initial standard for this book is that the earth rotates and stays in place, and the sun evolves around the earth. With the formulation that there is a change in the location of the longitude which is exemplified by 0 degrees longitude at 0.9856 degrees per day. With the sun also moving as much change or displacement of that position in one day. That is, the point of 0 degrees and the point of the sun's travel shifts by 0.9856 degrees to the east every day. That is to say, stars that are wholly or wholly behind the sun will be seen alternately. We will only see stars that are in the outer lines of the position of the sun and the earth which we can only look towards in front of the sun or those in the outline. We will not see stars that are behind the sun because starlight will not be able to beat the sunlight.

Now we just example that in the sky there are 12 stars, each of which is 1/12 of the entire path of the sun. That is, the stars are always in place will look alternately and sequentially. In fact, the appearance of stars always shifts by the amount of sun shift and the shifting of the earth's longitude every day. However, by contrast because each distance is 1/12 of the great circle of the sun's orbit, you will see a change in the appearance of stars that are very different every 1 month or every sun completes 1/12 of its revolution. And with a distance of 1/12th of that star will be seen 6 stars each night that the contrast looks changed after 1 month. Taking for example the first day we will see stars 1 to 6 stars, which slowly each day the appearance of the star's position will change by 0.9856 degrees every night. After 30 days or 1/12 of the sun is running, 7 stars will appear and 1 stars will disappear from view.

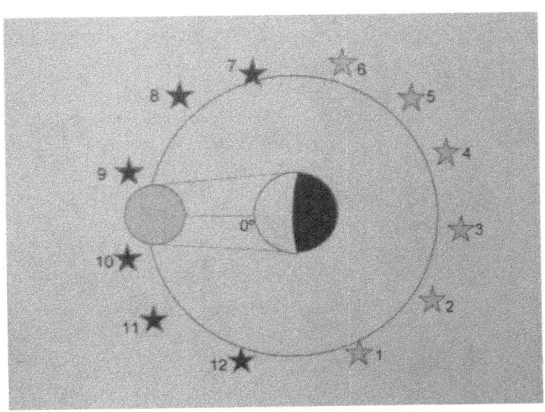

Pictures of constellations appearing on the First Moon, which are visible from 1 to 6 stars, we can see that the stars seen are stars facing the sun, while stars 7 to 12 behind the sun are not visible.

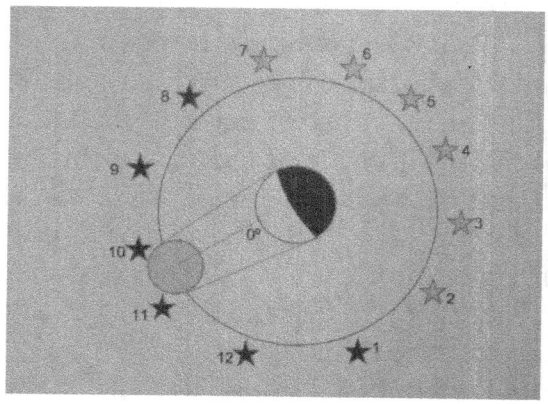

The appearance of constellations in the Second Month, what appears are stars 2 to 7, while 1 star disappears, 7 stars appear. The 0 degree Longitude has shifted 60 degrees to the east.

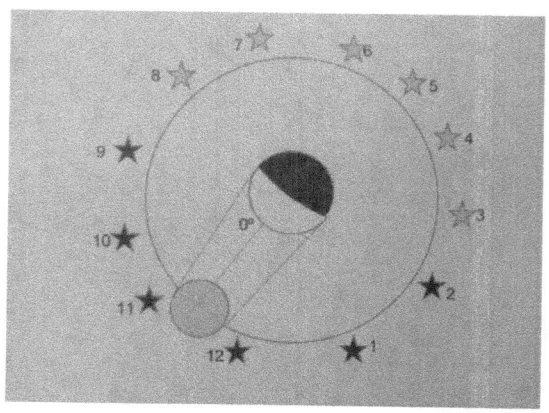

The appearance of constellations in the Third Moon, which is seen from 3 to 8 stars, while 2 stars disappear, 8 stars appear. The 0 degree Longitude has shifted 60 degrees to the east.

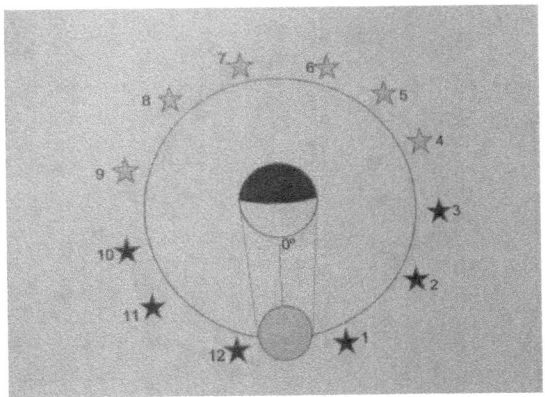

The appearance of constellations in the Fourth Month, seen from 4 to 9 stars, while 3 stars disappear, 9 stars appear. The 0 degree Longitude has shifted 90 degrees to the east.

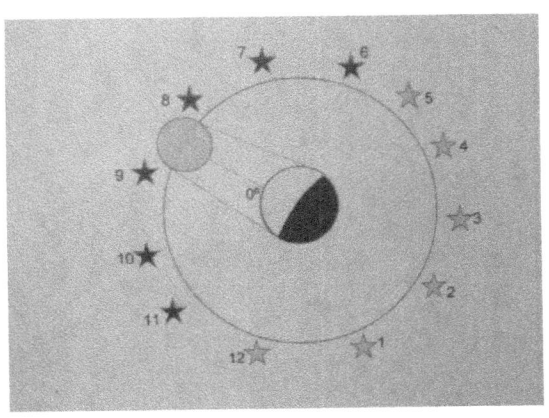

Image of constellations appearing on the 12th Moon, visible from 12 to 5 stars, which continue to occur regularly until longitude is 0 degrees and the position of the sun has returned to its original position.

SOLAR ECLIPSE

Eclipse is another natural phenomenon that occurs due to the relationship of the earth, sun and moon. Eclipses can occur when the moon is right between the sun and the earth, or the earth is between the sun and the moon. If the moon is between the sun and earth, a solar eclipse will occur. And if the earth is between the sun and moon, there will be a lunar eclipse. However, in a solar eclipse there will not always be a solar eclipse under these conditions, a solar eclipse will only occur if the moon is at the right angle and makes it possible to prevent the sun from being seen in any place or region on earth, either in part or in full. . Because we know that both of them have a certain degree of slope. Unfortunately, the author has not yet had time to measure the degree of slope of the moon's orbit towards the earth. So, be happy and I congratulate those who have succeeded in doing that. Because by knowing the magnitude of the degree of slope of the paths of the two objects and by calculating the speed of the sun and moon in evolution will make a person able to make an estimate - which, although not always accurate, when a solar eclipse or a lunar eclipse occurs. Congratulations interpreted here is a sign of respect for the struggle and diligence in researching because it requires a lot of energy, time, money and equipment.

A solar eclipse is not something special, and will not affect the wheel of life on earth. However, because an eclipse is something that rarely happens, making many people enthusiastic and want to witness it directly. Many people are even willing to go to a place far away and require no small amount of money just to see an eclipse. Moreover, a total solar eclipse, which is because of its rarity has made it like a pearl that must be obtained.

By observing the movements of the sun and moon, also without using any measuring instrument, we can draw the conclusion that the moon's tilt is indeed greater than the sun. Because we witness every month there must be a dead moon. But even though the location of the moon at that time was right between the sun and the earth, there was not always a solar eclipse. Yes, throughout the surface of the earth. The sun looks normal and nothing strange happens. Here I mean the eclipse of the sun is not the location where the moon's shadow is, but seen from the area where the sun is covered by the moon. If in the sense that during this solar eclipse is interpreted as a place where the moon's shadow falls, but I prefer to call it as a place where the sun is covered by the surface of the moon. Although the consequences are the same, they are in a different meaning. This means that someone will see a solar eclipse not because it is in the shadow area of the moon, but because his view of the sun is covered by the surface of the moon. Because if it refers to the shadow of the moon there will be no such thing as a partial solar eclipse, because the area experiencing a partial eclipse does not see the darkness of the sky but because our view of the sun is partially covered by the moon. It is rather difficult to understand or be accepted by reason and logic, but indeed this is the case. That a person does not need to be in the shadow of the moon to see a solar eclipse, but rather be in any location on earth that allows to be able to see how much the surface of the moon covers some or all of the disk of sunlight. With the more precisely a person is in the right location too, he will see more and more of the sun's surface covered by the moon, because despite the occurrence of a solar eclipse, not all people on the entire surface of the earth who in the afternoon can see the eclipse.

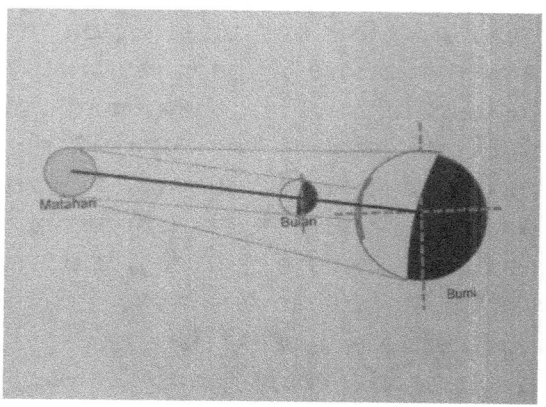

Drawing at a glance sketch of a solar eclipse

In the picture above, it can be explained that a solar eclipse occurs in an area that allows to see the globular moon covering the sun. So, the orientation or yardstick is not a shadow. The benchmark is how precisely the moon disk covers the sun disc. In areas with red stripes the sun will be completely covered without considering what the actual size of the sun is. But what is clearly seen is that the sun and moon look the same or the same size, so that the moon covers the entire sun. At this location a total solar eclipse occurs. Whereas the green stripe allows the sun to be seen from the area partially covered by the moon. Once again, the benchmarks are not shadows. However, this condition can only last a few moments in a very short time, because it depends on the speed of rotation of the earth, the speed of the revolution of the moon and the sun, and how the degree of rising or falling of the tilt of the moon and the sun against the earth.

GEOCENTRIC AND LOGIC

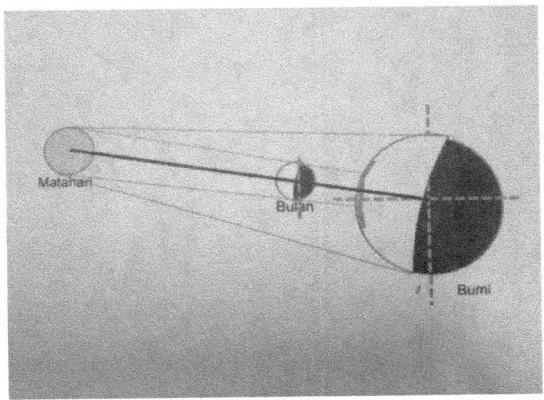

*Draw one of the moon's tilted position
which causes a solar eclipse*

The black line in the image above is the focal point of the sun's irradiation of the earth, and the two red outermost lines are the entire solar irradiation area. This we assume the occurrence of a dead moon or new moon when the sun is at the peak point of the slope of its orbit. If the moon is still in the area which is the outermost limit of solar radiation, then it is certain that a solar eclipse will occur in one area or part of the earth's surface even though it is only visible from the northernmost or southernmost regions of the earth. So to avoid this from happening, the moon must come out of the outer boundary line. And by coming out of the area of the sun's irradiation of the earth will make the moon will not obstruct our eye's view of the sun.

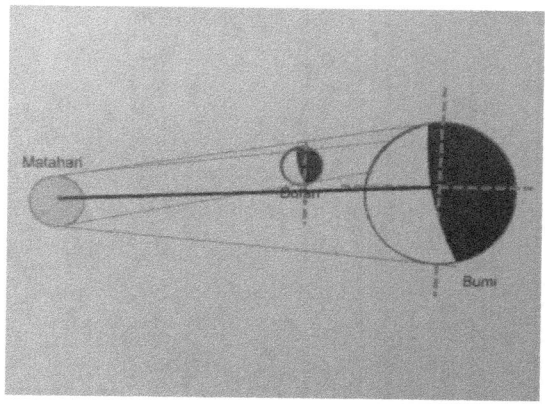

Figure Another position of the moon's tilt that causes a solar eclipse

So it is clear that with the two pictures above, it can be concluded that there will always be a solar eclipse at the time of the dead moon or new moon if the moon does not come out of the sun's exposure to the earth's surface, because the moon can still obstruct our view of the sun on any of the regions on the surface earth.

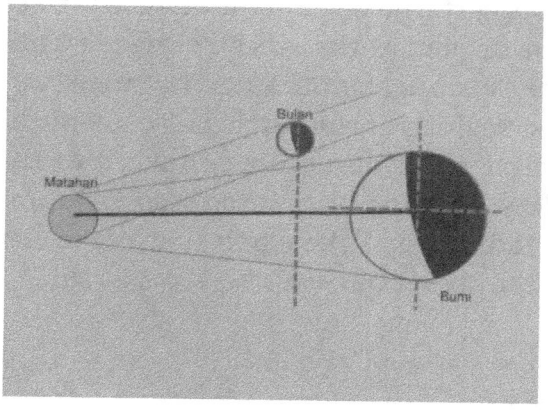

Figure The position of the moon's tilted line to prevent a solar eclipse.

By looking at the picture above, we can see that the moon has come out of the outer limit of the sun's irradiation of the earth. With this position the moon will not be able to obstruct our view of the entire earth's surface of the sun's light or irradiation so that no solar eclipse occurs. So it can be concluded that the slope of the moon's path must be greater than the slope of the sun's path.

LUNAR ECLIPSE

As the occurrence of a solar eclipse, the lunar eclipse process is almost the same. The moon will experience an eclipse when the earth covers the sun's light so it is not visible from the surface of the moon, one difference is the location or position of the three objects. If when the solar eclipse of the moon is between the sun and the earth, then when the lunar eclipse occurs the earth is between the moon and the sun. Lunar eclipses also occur as a result of their reactions or movements and the movement of the sun. And in fact the concept of a lunar eclipse is the same as a solar eclipse. If we call a solar eclipse is because our view of the sun is covered by the moon, then during a lunar eclipse will also be like that if we are on the moon. Our view of the sun if we stand on the moon will be covered by the sphere of the earth. So the concept is the same. However, the notion of a solar eclipse and a lunar eclipse have little difference. The difference is that when a solar eclipse occurs, it can be called darkness covering the light source. Namely if it were not for the globular moon covering the sun the eclipse would not have occurred.

The difference between the mention of a solar eclipse and a lunar eclipse is because we are on earth. So if we are on the moon, we will also call it a solar eclipse. With the same concept that we cannot see the sun from the moon because the sun is covered by the sphere of the earth. So because the earth and the moon are the same nature that does not have its own light, and must get sunlight to look bright. So because we are on earth, we call it a lunar eclipse. Another explanation is that because we are on earth a lunar eclipse occurs because the shadow of the earth covers the sun's light so it does not reach the moon.

It is a different concept if we look at it from where we stand now, if we experience a solar eclipse it is because it is not hit by shadows. However, because the sun is covered by the moon, so the appearance of the eclipse is very obvious. So when we call a lunar eclipse, it would be true that the sun's light did not reach the moon because it was covered by the earth or the shadow of the earth fell to the surface of the moon which was supposed to get light. So the explanation is short, because the sun has light that spreads and breaks down, if the moon does not cover the sun's circle there will not be a solar eclipse. Whereas during a lunar eclipse the earth falls to the moon, so the lunar eclipse becomes less noticeable or the boundary line is visible. Even when a lunar eclipse sometimes only changes in color and not always the surface of the moon looks dark or black, because the description or the distribution of sunlight to the surface of the moon cannot be completely covered by the earth or its shadow.

With the size of the earth larger than the moon, and the moon circulating around the earth, the moon does not have to be in a straight line or right in the middle of the earth to cause an eclipse. Because the concept of a solar eclipse is not oriented to the shadow, so that when the sun is in a certain position a solar eclipse will still occur even though the moon is not in line with the center of the earth. If we draw a conclusion, a solar eclipse will still occur even though the midpoints of the three objects can form a triangle. But in a lunar eclipse, the moon does have to go into the shadow of the earth for an eclipse to occur. Because with the sun's light that spreads and breaks down and its size is said to be too big compared to the earth and moon, the shadow will shrink. Unlike a small light source, if the light source is small then the shadow of an object will be larger than its actual size.

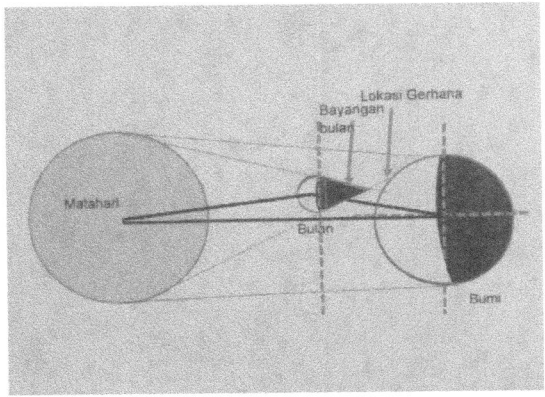

Image of example position of the moon during a solar eclipse.

If we look at the picture or diagram above, the moon's shadow does not seem to reach the earth. But there is still an eclipse at the location indicated by the arrow because our view of the sun at that location is covered by the moon. So that we need not hesitate to mention that a solar eclipse is not caused by the fall of the moon's shadow at a place or a location on the surface of the earth. But it is caused by obstruction of our view of the sun because it is covered by the globular moon.

This is the difference with a lunar eclipse. For a lunar eclipse to occur, the moon must enter the shadow of the earth. With a size smaller than the earth, causing the moon can or may enter the shadow of the earth and stay longer in the shadow.

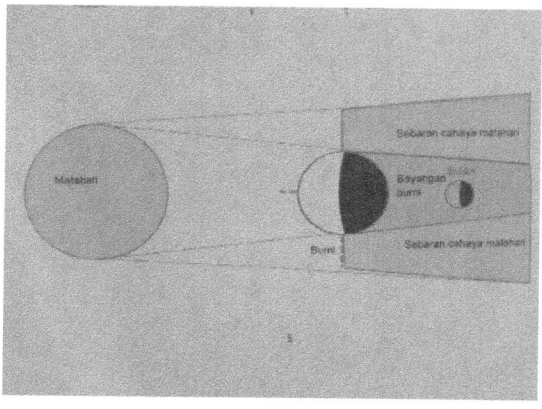

Lunar eclipse

By looking at the picture above, we can see the shadow of the earth falling to the surface of the moon is not too dark due to the still spread of sunlight. Unlike solar eclipses such as total solar eclipses, the entire surface of the sun will be covered by the disk of the moon. Meanwhile, the distribution of sunlight during a lunar eclipse causes the earth's shadow to not be very dark, so that the moon is still visible when an eclipse even though it only looks faintly or changes color, because the distribution of sunlight is always the same in all directions around it.

The time of the lunar eclipse is the opposite of the solar eclipse, if the solar eclipse generally occurs when the new moon, then the lunar eclipse occurs when the moon enters the highest phase, or during the middle of the Islamic month. This means that it can be on the 13th, 14th, or 15th day.

SEASON

In general, the season or the change of seasons is caused by the impact of the heat of sunlight and the magnitude of the degree of the sun's slope towards the equator. The sun that circulates continuously with this constantly changing slope will sometimes make it right at the equator and move to the northern or southern hemisphere. With his regular trips resulting in regular seasonal changes on the surface of the earth. Seasonal changes are particularly noticeable in regions or regions on the earth that are north or south of the equator. If at the equator only two seasons are known, namely summer and rainy season, although sometimes erratically, in the north and south of the earth there are four seasons each year. Namely summer, autumn, winter and spring. And with research and attention to natural phenomena, humans have been able to predict and determine when the seasons will come.

The season has indeed become a natural factor that

inevitably must be followed by humans. And with that compulsion man must think how to deal with it. Because by being ready to face the changing seasons, humans have been able to prepare anything that needs to be provided to deal with these seasons. Because now we not only know air conditioners for cooling or air conditioning, but also have been created tools for heating the room. Its use is certainly different, if the cooler is certainly used when the weather is hot and when the weather is cold, naturally heating is needed. And in the past people have used coal burned as heating or conditioning the room in winter.

In the case of the change of seasons which is influenced by the position of the sun against the earth, the opposite occurs between the northern and southern hemisphere earth. If the northern hemisphere experiences summer, the southern hemisphere will experience winter. Vice versa, when the northern hemisphere experiences autumn, the southern hemisphere will enter spring. The change or change of this season has become a habit or a common thing for humans since ancient times, as well as anything that needs to be prepared to face a certain season.

The sun is right on the equator twice a year, on March 21 and September 23. March 21 is the beginning of spring in the northern hemisphere and the beginning of autumn in the southern hemisphere. The sun will continue to rise to the north until it reaches its peak on June 22 or 23, within a span of three months this is spring in the north and autumn in the southern hemisphere. The four seasons on average did last for three months each. March 21 to June 22 is spring in the northern hemisphere and autumn in the south, June 23 to September 23 is summer in the northern hemisphere while in the southern hemisphere is winter. On September 23 to mid-December is autumn in the northern hemisphere while the southern part experiences spring, and from December 22 to March 21 is winter in the north and summer in the south. Do we need to ask or not, why is not the beginning of the season which is used as the beginning of the Islamic calendar because it will be more measurable and certain. But it might be caused by other factors that we do not know what is used as a reason to make January 1 as the new year BC.

SPACE AND TIME

Space and time are two inseparable things. We live indoors. The relationship with time is the smaller the room, the less time will be needed to get to know and explore the room. But humans have wide eyes and distant vision. The farther a human's perspective is, the more he will know that he lives in a very large room, and he increasingly knows that humans need even more time to know what is inside the room and how to explore and even master it. With a sense of curiosity and a sense of wanting to conquer or master, humans continue to try to explore the contents of his head how to do that. We might think that space has limits, but how to translate time is very difficult to think about. So we feel that we will be able to control that space but we should be able to do it in a limited amount of time. But until now humans must admit that we have not been able to overcome or control the space because of the limited time.

GEOCENTRIC AND LOGIC

a normal instinct that happened to him because he felt he had done the right thing. What's worse is the person who closes his heart to all the whispers of kindness, so that no matter how bad he does he will still feel right or not feel guilty.

The whole world recognizes that, all the worlds that are interpreted are all people, they all always try to make things better. All rules or laws that are made are the result of the engraving of people who feel that what is enacted will make the world travel and travel wheels better. Although whatever name may be called ideology or ideology, everything is made and arranged with the aim of making things better, both for himself and his nation or class. And with the large number of people will make a lot of thinking brain or head, which as we are witnessing now makes various types of life holdings for humans both in terms of personality or nationality. All of that is a normal thing to happen, because humans each walk and start their steps in accordance with the guidance of the heart and mind. Sometimes what he conveys is a guide to the lives of millions of people and even becomes a path that must be followed by everyone who can receive the results and contributions from his thoughts.

But unfortunately, of the many ideas put forward and have become the handle of the lives of many people who are considered right by who sparked it, instead bringing all those people into darkness. Because it is undeniable that with instinct and mentality people always try to make their lives better, even though sometimes they have to sacrifice a large number of others. But with the instinct of humanity that always feels like to be the best or the strongest will make humans go all the way so that their desires are achieved. It is not wrong for humans to think like that, because he knows there are whispers in his heart and soul that must be followed. But most people prefer whispers that according to themselves are good even though everyone thinks it is wrong. The form of his mistake in obeying the whispers in his heart is that he will get a bad reply from what he did. Even sometimes have to be willing to live and their bodies separated due to the misinterpretation of the whispers in the heart.

A human being does need guidance and guidance, and is influenced by relationships and what he knows. The biggest thing that must be sorted out by humans is a whisper in their own hearts. That whisper will determine his way of life and daily life. If a person chooses the two whispers in his heart it will lead him astray. And if the only ones who get lost are probably not to have too much impact on others. What is frightening is that many people who for whatever reason also want to participate in the error, because of the error they consider also good for themselves.

DEITY

Happiness and suffering are indeed measured by each human person. The feeling of happiness and suffering depends on the person who feels it. A person may feel happy when what he wants is achieved, or he succeeds through suffering and that suffering ends. Or a parent feels happy and happy to see their children can achieve stability in life, and vice versa when a child is able to fulfill the wishes of his parents. And many other parables. But in the modern era and completely free as it is today, do you know who is happiest?

The happiest people are those who feel themselves free and don't need to think about bonds or a law that is blocking their path. He feels free to do whatever he likes without thinking about the consequences of his actions. And as we know and we must admit is that the bond or the barrier of humans in doing and so that they cannot live as they wish is religion or belief in the existence of God. God sent his messenger to invite and show people to the right path and good. If someone acknowledges that there is a god and justifies what is taught and delivered by the envoys, then he must live in signs that will directly bind him and not allow the believer to do as he pleases.

There are not a few people who choose to live freely without thinking about religious ties, the basis of which is God. The freedom that is felt is extraordinary satisfaction and happiness. By living freely humans will feel that the world is very roomy. And as we know, the idea of the absence of God has been spread and campaigned by people since berates years ago. Many people who believe and go with the flow are caused by a feeling of wanting to be separated from a bond, wanting to live freely and at will. Although in some human-made law they are still bound, but in his heart there is a sense of satisfaction that is held to hold that whatever he does will only have worldly effects, so he does not need to think about how much he did wrong because he knows the rewards of what he did it is only physical without any calculation after he dies. It is these people who have the opportunity to master all the fairies of life in this world because they do not know whether something is permissible or not, and whether something is bad or good.

Belief in a god will make a person think more about how he produces good results from his actions, which can be interpreted outwardly as how

God will be happy to him and give a good reply. And for people whose sense of divinity is already high or above average, it can make it no longer too undecided about what will happen to him in the days to come. Because he knows that while he is doing good his good will bring good to him.

It's amazing, someone is able to think that everything that exists and everything that happens is by itself, there is nothing to regulate or supervise. Especially to the extent of reasonableness in thinking and conscience, man is able and courageous to assume that he can empty his heart and mind about the existence of an extraordinary power that controls at least his own conscience. He dared to lie that in his heart there was a flash of whispers that directed his steps. And he too knows that the whisper took him to a very decisive way towards the end of his journey. It has been said that if only he himself goes and chooses the path which is said to be true happiness, then it does not have a great effect on others. But unfortunately many people who disagree and finally take the path he shows. And more unfortunately, what he taught has become a necessity in the management of human relations, although sometimes forced.

But, that's life. We can only do according to our abilities, and maybe we are people who believe in God but are in faith as weak as faith. Only able to hate what other people do without being able to change or stop that person from doing wrong. But, the wheel of life continues to spin. Maybe someday we will achieve what we imagine and end up as expected. But no matter how hard we try to do good to others with the aim of inviting them to the good too, in the end it is God who decides. Because God himself has stated clearly and clearly that those whose hearts are diseased, will be locked to death. So that he will no longer be able to accept the truth. And we can only be grateful for the blessings that we can immediately feel, even though we are not unconcerned, but because someone's heart has been locked dead by the one who created that heart, then no one will be able to open it again. Because how evil a man is if he still has a sense that he will return to the day he has to face and feels responsible for what he does, maybe that heart can still be opened with the clarity of the soul.

It is sad enough to hear the statement that religion, which means God, is a barrier to progress. The presumption expressed as if ignoring the importance of religion for human life. In general people who think like that are due to hearing and reading or following wrong. What does it mean? Beliefs or religions that prevent people from thinking or being knowledgeable are false beliefs. Or maybe even people who understand the problem of religion are making lies or lying to people. Because not all religions forbid humans to achieve progress or success, both in the material or knowledge fields. And these teachings even demand that humans continue to

GEOCENTRIC AND LOGIC

learn. Not just learning about world life, but also for success or happiness in the hereafter.

However, human beings who have been feeling depressed and constrained by religious ties have already felt themselves right, that they will be able to make changes to human life to be better without the existence of divine values. And they also feel victorious when they can escape from the influence and religious norms, by proving that they can do more for the advancement of science. And we can clearly say that it may not be the teachings of the religion that are wrong, but because the religious leaders themselves feel that they should not be wrong and should not be opposed even if the religious leaders know that what he is conveying is very contrary to the teachings of his own religion.

It might even be said or concluded that the leaders of a religion who always want to monopolize, so that makes many people feel and make decisions there is no point in divinity in life. Because if they are religious they will be bound and constrained by all their actions, both words and deeds. So it can be said, one's decision to choose no religion is caused by the delivery of false teachings from the religious leaders themselves. So that someone who removes the divine values in his life feels more able to do something that he said is good for the progress of all humanity rather than following religious teachings that curb freedom. But I explained that both deeds and opinions were wrong, people who taught lies in religion were clearly wrong. Whereas those who choose not to have a religion or believe in God are also guilty. Because if someone knows that what is taught by his religion is wrong it should not come out of its nature as a godly being, because there are still other religions that can be chosen as a way of life that teaches and requires its adherents to continue to learn and progress in all fields.

God is indeed too difficult to understand and accept by people who are accustomed to living freely, without straps blocking their steps. But they may not yet realize that religion is a guide, not a binding. Guide means the right direction. That if someone who has knowledge will be more directed in applying the knowledge and intelligence of his brain for good to the path of truth. In contrast to people who are knowledgeable but do not have confidence in the Godhead, he will think and do whatever he wishes with the knowledge he has. It does not matter whether it will produce good or bad. People who do not have a religion will feel nothing will happen to him even though whatever he does, because he does not realize and does not know or even does not acknowledge that every step in every breath and heartbeat will be held accountable. Therefore, if you are someone who is given advantages in terms of science and intelligence, use that knowledge in accordance with the teachings of truth and not to produce evil. Because if you do good then good is for yourself, and if you do bad then bad is also for yourself.

Studying knowledge is an obligation so that we avoid ignorance and backwardness. But demand knowledge that is appropriate and obedient to the teachings of God and does not run from the path of truth. Because, no matter how high knowledge is meaningless without divinity. And it's useless claiming to be godless without having knowledge. Because there is no knowledge that does not originate or without guidance from God, and there is not a single true religious teaching that does not lead to the right path. And to find a truth we must have knowledge.

Salam...

Leo & Knowles Publishing

THE COMPLETE BEGINNER'S BLUEPRINT TO STOCK MARKET TRADING

The Complete Guide

CANDLESTICKS + BREAKOUTS PATTERNS + INDICATORS + ENTRY EXIT + RISK MANAGEMENT

© THE COMPLETE BEGINNER'S BLUEPRINT TO STOCK MARKET TRADING

The Complete Guide
CANDLESTICKS + BREAKOUTS PATTERNS + INDICATORS + ENTRY EXIT + RISK MANAGEMENT

Published by: Leo & Knowles

www.leoandknowles.com
Sales@leoandknowles.com

New Delhi-110002, INDIA
Order Online or by email.

Cover Design by Leo & Knowles
Art Work by Leo & Knowles

2024

All rights reserved. No part of the book or part thereof, including the title of the book, be reprinted in any form or language without the written permission of the author and the publishers. Any infringement of the Act shall be prosecuted. Typesetting, fonts, and book cover design copyright to the publisher.